CAREERS IN THE
UNITED STATES AIR FORCE

CHOOSING A CAREER IS TOO IMPORTANT to put off until the last minute. Even while you are still in high school, you should be thinking seriously about what you want to do with the rest of your life. Your school years will prepare you for a productive career. You owe it to yourself to give this process the serious thought it deserves.

A career is about more than dollars and cents. There is a "big picture" to keep in mind when homing in on what you would like to do when you get out of school. Everybody wants to do something they enjoy. Most people also want to do something that they find meaningful. Most people want to become part of something larger than themselves, too.

A career in the United States Air Force can provide you with all of these things. You do not have to be a pilot, either. In fact, relatively few Air Force personnel fly planes. Like all branches of the US military, the Air Force offers hundreds of individual career paths ranging from intelligence analysts and air-traffic controllers to chefs, as well as pilots. The Air Force is a great place to learn a skill and get a start in life – or to spend a 20-year career. You will have that choice if you enlist in the Air Force.

The US Air Force is the largest and most advanced air force in the world. It counts 328,000 personnel on active duty, 74,000 in the Air Force Reserve and 106,000 in the Air National Guard. Those personnel are responsible for nearly 5,800 aircraft ranging from tiny trainers to long-haul bombers to super-sophisticated fighter planes. This number is larger than the next two largest air forces – Russia and China – combined. The Air Force is also responsible for the United States' arsenal of intercontinental ballistic missiles, is the lead agency on launching military satellites and owns a rapidly growing number of unmanned combat aerial vehicles. Simply put, the US Air Force has no peer – not even close.

Air power is an important part of the comprehensive package of US military might. Ground, sea and air power all play a part in winning wars and maintaining the peace. Ever since the dawn of air power it has been understood that the first step in controlling the ground is controlling the air. Since the goal of any war is controlling the ground, a modern war cannot be won without air power.

Take careful note of the information contained in this report. In it you will find sections on how to prepare for your career, what kind of education you will need, what you can expect to like and dislike about the career and even how much money you can expect to earn at various stages along the way. If you like what you read here, talk to a recruiter. Remember, just talking to a recruiter incurs no obligation on your part. Requirements and opportunities change constantly.

You can be learning about a career in the Air Force even while you are still in high school. If your school offers a Junior Reserve Officer Training Corps program (JROTC), join it. Many high schools sponsor JROTC programs. Similar to the college ROTC program but without the legal obligation to serve after you graduate, JROTC is a great way to learn about the military. Specific branches of the military cosponsor JROTC programs. If your school's JROTC program isn't sponsored by the Air Force, join it anyway. You will learn a lot about military bearing and history, and get to spend weekends doing very cool things that most teenagers can only dream about. You will be issued a great looking uniform, and you will get to wear it to school once a week.

If your school does not sponsor a JROTC program you can still prepare yourself for an Air Force career through more conventional means. The most important thing you can do is to study hard. The modern US military is a very sophisticated operation staffed entirely by volunteers. Those volunteers do not want to put their lives in the hands of someone who is not prepared intellectually. Develop good study habits. Not only will you get smarter, you will also prepare yourself for the constant training and education you will undergo after you join. Work equally hard in gym class, too. The military's physical-fitness requirements are not superhuman but they are challenging. Couch potatoes will not get very far.

There is no lack of easily accessible information about the Air Force. Browse the web and search the stacks at your school library. Subscribe to Air Force Times, a privately owned newspaper that covers the Air Force. There is certainly no shortage of movies with an air power theme. Watch as many as you can. They may or may not be very accurate but they will keep your enthusiasm running high, and that is important too.

HISTORY OF THE CAREER

FOUNDED AS AN INDEPENDENT SERVICE IN 1947, the Air Force is by far the youngest of the military services. The Constitution specifically gives Congress the responsibility to "make rules for the government and regulation of the land and naval forces." It also makes the President the commander-in-chief of the Army and Navy. Written in 1787 and ratified in 1788, the Constitution does not mention an air force because the Founding Fathers never anticipated one. The first airplane did not take to the skies until 1903.

It took longer for the military to express interest in airplanes than you might think. The earliest aircraft, like the Wright Flyer that famously flew at Kitty Hawk, North Carolina on December 17, 1903, were flimsy contraptions that stayed aloft almost by accident. They were slow, difficult to maneuver and could not carry large loads. The one thing they could do, however, was take a pilot high above enemy lines where he could spy on hostile formations. The Army had begun experimenting with using balloons for overhead reconnaissance in 1907. It added airplanes to its arsenal in 1908.

Aviation technology and training took five years to be recognized in the US military, and the 1st Aero Squadron was formed in 1913, becoming the first US military unit exclusively devoted to flying. When World War I broke out in Europe in 1914, the squadron consisted of six aircraft. By the end of the war in 1918 it had grown to more than 700 planes. This number still represented less than 10 percent of the total number of aircraft operated by allied militaries. The major European countries, especially Germany, France, Italy and the United Kingdom, had all pursued military air power much more aggressively than the United States.

As militaries tend to do after wars, the Army and its aero squadrons downsized dramatically. In 1920, the squadrons were reorganized into the Air Service, a combat arm within the Army commanded by a two-star general. This name was changed to Air Corps in 1926. The Corps grew rapidly in the late 1930s when World War II loomed on the horizon, with Congress authorizing $300 million for 6,000 planes.

The Army Air Corps was deployed around the world, fighting in the European and Pacific theaters. Aerospace technology took one giant leap after another, as the fabric-covered airplanes of the 1930s gave way to all-steel fighter planes and long-range bombers with enormous capacity. Some of the tactical aircraft designed for the war, especially the American Boeing P-51D Mustang, British Supermarine Spitfire and Japanese Mitsubishi Zero, were so capable that they are still in use today for aerobatics and racing. By the end of the war in 1945, the Air Corps had grown to 63,715 aircraft, a breathtaking leap from 2,200 in 1939, and strong testimony to the prowess of American industry.

Air power handily proved itself during the war, prompting the government to finally follow the lead of other major countries and turn the Army Air Corps into an independent service in its own right. The United States Air Force was born on July 26, 1947 with the passage of the National Security Act of 1947.

The modern Air Force grew up during the Cold War that followed World War II. Tensions between the United States and its allies in Western Europe, and the Soviet Union and its client states in Eastern Europe, meant that the military did not shrink as much as it usually had after a war. American forces had to be ready to fight World War III against a numerically superior foe.

The US military sought to level the playing field with technology. Space-based systems, advanced communications and nuclear weapons all played a role in ensuring that the US and its allies would be prepared. The Air Force benefitted from this phenomenon more than any other service. Air power had played a critical role in the war. Most Americans had never been on a plane, making aerospace technology exciting and new. Look at photos of the automobiles of the 1950s and 1960s. Tail fins, taillights shaped like jet exhausts, hood ornaments that looked like airplanes. Even trains were fashioned to look "aerodynamic." Americans were fascinated by air power and were only too happy to pay for the world's greatest air force.

Although the development of jet fighters often generates the most interest, advances in delivering firepower did the most to change the way in which wars are fought. The first operational jet aircraft, the F-86 Sabre, was introduced in 1947. Jets were faster and more reliable than their propeller-driven predecessors. The mammoth B-52 bomber made its debut in the 1950s and is still in use today. With extremely long range and the ability to carry tons of bombs, the B-52 could deliver multiple conventional bombs or one nuclear bomb to almost any point on the planet. The development of missiles pushed the envelope one step further by allowing massive firepower to be delivered at no risk: no pilots needed to fly into harm's way. Stealth technology developed during the 1980s allows aircraft to penetrate deeply into enemy airspace while remaining invisible to radar, increasing lethal potential while reducing risk.

The key to the success of the Air Force during the Cold War is a strategic phenomenon known as force multiplication. One bomber with a crew of four can deliver a nuclear bomb with enough firepower to wipe out an entire city. A missile with a nuclear warhead can accomplish the same feat with no crew and no risk to the side that pushes the button. Soldiers, by comparison, can generally only shoot one enemy soldier at a time and have to put themselves at risk to do so. This is a staggering development in the history of warfare and has rightly generated enormous controversy. At the time, however, advanced technology was the most effective and least-expensive way for the United States and its allies to stand up to the huge conventional armies of the Eastern Bloc.

The post-Cold War Air Force is a technological masterpiece. Substantially smaller than in its Cold War heyday, the US Air Force can deliver more firepower than ever before. From their home base in Missouri, B-2 Stealth bombers can deliver guided munitions to any spot on the globe. Satellites launched by the Air Force can peer into places no other set of eyes can. Air Force intercontinental ballistic missiles are aimed at . . . somewhere. The fact that the targets are secret does a great deal to keep the peace.

The Air Force is also the world's preeminent provider of humanitarian assistance. In fact, most of the world's humanitarian assistance that does not travel by sea gets to where it is going compliments of the US Air Force. No other air wing in the world owns as many cargo aircraft able to go as many places. The cargo workhorse, the venerable C-130, can land almost anywhere and has been used to bring food, medical supplies and relief workers to every corner of the world. You don't have to watch TV news for very long to see a USAF C-130 in the background. If you want to see the world and do some good while you are at it, the Air Force may be your perfect career choice.

WHERE YOU WILL WORK

ALL US MILITARY PERSONNEL CHANGE jobs every two to three years, and Air Force personnel are no exception. Most job changes also come with a move, known in military jargon as a "permanent change of station," or PCS. Personnel are sometimes able to string together two jobs in the same place but three or more is very rare. Personnel are usually asked where they would like to go, and are given a few options. The Air Force extends options when possible because doing so keeps people satisfied, and they tend to stay in the service longer. However, the needs of the service will always come first. Ultimately you will go where the Air Force needs you to go.

The Air Force maintains hundreds of major and minor bases all over the world, including Air National Guard bases. The nature of air power requires that aircraft have access to as many runways as possible. Often, runways and fuel are all that are really needed in a particular place, so many Air Force bases are very small, consisting of a runway, refueling facilities and a few hangars. Bases that fit this description can be found all over the world. In some places overseas, the Air Force does not even have its own bases, but stations a handful of personnel on foreign air force bases or even at commercial airports to service USAF planes. You could be assigned to any of these small installations.

You could also be assigned to a sprawling base like Andrews Air Force Base in Maryland, near Washington, DC. With 20,000 active duty personnel on 4,300 acres, Andrews is one of the largest installations in the US military. Major bases support their own grocery stores and department stores, known as commissaries and exchanges. They also boast their own hospitals, gymnasiums, restaurants and clubs, fast food establishments, bowling alleys, movie theaters, carwashes and a host of other services.

Most Air Force installations are located in the United States. The service also maintains a large network of overseas bases. Major Air Force installations outside the United States are located in Italy, Turkey, Japan, Germany, the United Kingdom, the Netherlands, South Korea, Portugal, Spain and Kyrgyzstan. Minor installations are located in Djibouti, Ecuador, Belgium, Canada and really remote places like Diego Garcia, Ascension Island and Antigua, among others.

Many people find the opportunity to live in exotic places to be one of the best benefits of military service. How else could you live for a few years in England, or Diego Garcia? While many people thrive on the adventure, some people burn out sooner than others. This is especially true for military personnel with spouses and children. Moving families around every few years can be very disruptive. You may or may not like this aspect of military life.

DESCRIPTION OF YOUR WORK DUTIES

THE AIR FORCE OFFERS HUNDREDS OF rewarding opportunities for enthusiastic young people whether you plan to pursue a long-term career or just stay in a for a hitch or two and get a head start in life. The Air Force obviously employs many pilots, but it also employs intelligence analysts, personnel specialists and chefs, among others. The best part is that the Air Force will train you to do whatever it needs you to do. If you stay in long enough you will be trained to do many different jobs, some of which would have never occurred to you.

Like all other military forces, the Air Force is divided into enlisted and officer communities. The legal difference between enlisted personnel and officers has deep historical roots going back to the days when aristocrats were knighted by kings and given the authority to enlist soldiers to fight wars. The system works essentially the same way today, although today's officer candidates no longer need royal blood to apply for a commission; a bachelor's degree will do.

The easiest way to sum up the difference between officers and enlisted personnel is that officers are leaders and enlisted personnel are doers. Only about 15 percent of Air Force personnel are commissioned officers. Their job is to make policy decisions and direct the efforts of enlisted personnel to meet those stated objectives. Officers have more authority than enlisted personnel, and are paid more, too. But they also have to assume legal responsibility for the actions of those under their command. Many officers have been fired due to bad decisions made by junior enlisted personnel they hardly knew.

Enlisted personnel are the technical specialists of the Air Force. They are the people who spend years becoming really good at doing specific jobs. Skilled enlisted personnel are the backbone of every military service, and the Air Force is no exception. For every pilot – all of whom are officers – there are at least eight enlisted personnel entrusted with maintaining the pilot's plane. If you really want to lead people, set your sights on becoming an officer. Don't let the authority and paycheck cloud your judgment, however. If becoming a highly skilled expert in something is what really attracts you, then enlist.

Air Force enlisted jobs are known by their Air Force Specialty Codes, or AFSCs. AFSCs are grouped into categories and career fields. Some career fields have only one AFSC while others have several. Officer jobs are grouped into utilization fields and specialties. Although the titles for categories and utilization fields are sometimes different, certain groups of officers and enlisted personnel always work together. Officers in the Pilot field, for example, will always work alongside enlisted personnel in the Aircrew Operations career field, which is part of the Operations category. The distinctions are confusing at first but will become second nature after you have been in for a while. In fact, after a few years you will learn to speak Air Force so well that you won't get it when your civilian friends have no idea what you're saying!

Enlisted Job Categories

Operations

Maintenance and Logistics

Support

Medical and Dental

Legal and Chaplain

Finance and Contracting

Special Investigations and Special Duty Assignments

Operations category includes these career fields:

Aircrew Operations

Command and Control Systems Operations

Intelligence

Aircrew Protection

Safety and Weather

Maintenance and Logistics includes:

Manned Aerospace Maintenance

Communications and Electronics

Fuels

Logistics Plans

Missile and Space Systems Maintenance

Precision Measurement Equipment Lab

Maintenance Management Systems

Supply

Transportation and Vehicle Maintenance

Munitions and Weapons

Support covers these career fields:

Information Management

Communications and Computer Systems

Historian

Services

Public Affairs

Security Forces

Civil Engineering

Mission Support

Manpower and Visual Information

Medical and Dental includes all medical and dental career fields.

Legal and Chaplain covers all fields related to law and religious programs.

Finance and Contracting includes financial fields and contracting careers.

Special Investigations is the home of the Air Force Office of Special Investigations.

Special Duty Assignments cover all functions that fall into any of the above categories and special projects outside the service member's normal duties.

Career fields are further broken out into specific AFSCs, not all of which are open to new accessions at any given time. Talk to a recruiter to get the latest news.

Officer Utilization Fields

Pilot

Navigator

Space

Missile and Command and Control

Intelligence

Weather

Operations Support

Logistics

Security Forces

Civil Engineer

Communications and Information

Services

Public Affairs

Mission Support

Manpower

Health Services

Biomedical Clinician

Biomedical Specialist

Physician

Surgery

Nurse

Dental

Aerospace

Medicine

Law

Chaplain

Scientific Research

Special Investigations

Special Duty

Like enlisted career fields, officer utilization fields are broken down into specialties. When you check with a recruiter to determine what fields may be open to you, also make sure to find out what prerequisites may be required for you to apply for a commission in a particular field. Some specialties, like civil engineering and all medical and legal fields, require candidates to possess specific degrees before they are allowed to apply.

AIR FORCE PERSONNEL TELL THEIR OWN STORIES

I Am a Fighter Pilot

"I have the career everybody thinks about when they think about the Air Force. The truth is that only about four percent of Air Force personnel are pilots and only about half of us are fighter pilots. But movies are made about fighter pilots, so we have the highest profile.

There's nothing more awesome than a fighter plane. I can't think of anything else that brings together so many technologies in one place and does it in such a devastatingly efficient way. Fighter planes are the embodiment of the highest technology. That may not be surprising, coming from me, but I've felt this way since I was in high school. That is why I set my sights on becoming a fighter pilot.

I was an Air Force ROTC cadet in college. The program paid for part of the cost of my education, and taught me how be an officer. I also earned a private pilot's license. I learned to fly on my own time and paid for the lessons myself, but I wanted to get started as quickly as I could. I also wanted to see if I was going to love flying as much as I thought I would – I did.

It takes more than loving to fly a plane, to become a fighter pilot, however. The requirements are incredible tough. If you have less-than-perfect eyesight, forget it. But that is just where it begins. Pilot trainees are subject to a battery of demanding tests that determine if they have the physical qualifications to make the cut. You'd be amazed how many people are dropped from the fighter program because of health problems they never knew they had. I had a friend, for example, who was dropped because of a heart problem. That sounds bad, but this particular heart problem only became apparent when my friend was subjected to enormous gravity forces only experienced by fighter pilots. My friend will go on to live a full and normal life but will never become a fighter pilot. People are merely moved into a different training program that puts less stress on the body, like flying bombers or cargo planes.

Initial flight training lasts a year. It is then followed by at least another year of specialized training. I fly F-16s, the Air Force's most-common air superiority fighter. We are never really out of training. We exercise against each other and in conjunction with Navy and Marine Corps air wings and foreign air forces. We simulate courses of action that could be taken by our adversaries around the world so we can be prepared for them if they ever happen in real life.

I've seen my share of combat deployments in my 10 years as a fighter pilot. When in combat you can never forget that the people on the ground want to kill you. That is especially difficult to do against the US Air Force because we have ways to mitigate any kind of air defense system. But we are not perfect. We can jam radars, bomb antiaircraft emplacements, fire chaff to fool missiles and beat anybody in the world in an old-fashioned dogfight. We can't avoid every danger, however. Once in a while one of us goes down behind enemy lines. Usually we get the pilot back, but not always.

I'm following the usual fighter pilot career path. I've done my training and many operational deployments. In the next few years I'll slide into positions in which I do more training and administration and less flying. That's the way it's supposed to work. Now that I have experience I can pass it on to up-and-coming pilots. Like most adults I now have a family to think about. I'm perfectly happy to take jobs that will give me more time at home. I still like the adventure as much as I ever did but I'm ready to turn it down a notch. I've already seen more adventure in the last decade than most people see in their entire lives."

I Am an Intelligence Officer

"I can't tell you exactly what I do. Most of the details are classified. There is misunderstanding out there, however, and I would like to clear it up. I am not James Bond. Intelligence is not espionage. Espionage – what James Bond does – is a field unto itself. It is very small and nowhere near as powerful as people think it is. It is related to intelligence but is a very, very small part of the big picture. Intelligence, simply defined, is whatever knowledge commanders need to make better decisions. Intelligence is information subject to professional analysis. That could be something as simple as taking the time to look up a weather report and make a recommendation on flying conditions, to something as complicated as learning about the air power available to a foreign leader in order to make a predictive assessment about what that leader will do if we attack from the air. Intelligence is classified simply because we don't want our adversaries to know what we know.

I am the professional in "professional analysis." I am one of the thousands of people in the Air Force who spend days sifting through mountains of raw information looking for the bits that might be important. Sometimes several bits that do not look like much separately become important intelligence when they're put together the right way. Finding these patterns and recognizing their significance are the most difficult things we do.

I've dealt with many types of intelligence during my career. The two I like best are imagery and open-source. Imagery intelligence is anything that involves a photograph, and it's something the Air Force does very well. We can take pictures from piloted aircraft, satellites, and unmanned aircraft that can fly low without risking a pilot. Imagery presents views of the world you can't get anywhere else. It's a challenge to find sites of interest on images that may cover many square miles. The first clue is to look for straight lines. Nature doesn't do straight lines. If I see a straight line I'll recommend we take a closer look – it's almost always something man-made.

I also enjoy open-source intelligence. The goal of open-source intelligence, as the name implies, is to glean intelligence from the vast amount of information flowing through open sources all the time. Newspapers, magazines, academic publications, speeches, television and radio broadcasts and, of course, the Internet are all sources of open-source intelligence. It used to be said that we could get 80 percent of the intelligence we needed from open sources. I'd say it's more like 90 percent today.

In either case, the idea is to sift through the mass of information and reduce it to the essentials the commander needs to make better decisions. My staff and I can spend all day poring over thousands of individual pieces of information and reduce it all to a single page of intelligence for our boss. That's professional analysis at work. It's not easy and most people aren't cut out for it. Intelligence professionals have every personality quirk imaginable, but they are all smart.

We are not all knowing, however. It ticks me off when the intelligence community is accused of making mistakes in war or in not magically predicting something like the terrorist attacks of September 11, 2001. Not only is there a vast amount of information out there, we are also subject to counterintelligence efforts designed by our adversaries to confuse us. Sometimes they work. All we can do is our best.

As an officer I am also responsible for leading a staff of intelligence specialists. Management comes with its share of headaches and triumphs, but I really like managing projects and leading people. I get a great sense of accomplishment when my team and I successfully answer the boss's questions, on time and in full. I wouldn't want to be doing anything else."

I Am an Airborne Battle Management Systems Sergeant

"I am an airborne battle management systems sergeant. I know, you're thinking that sounds cool even though you don't know what it means. Trust me, my job is really, really cool.

I run the sophisticated sensors and electronic countermeasures equipment aboard E-3 and EC-130 aircraft. These aircraft are flying computer platforms that carry our electronic eyes in the sky and the communications equipment necessary to link battle space nodes to one another. The battle space is what used to be called a battlefield, but includes assets and adversaries in the sky and, sometimes, at sea. A node is an individual asset, like a plane, tank or ship. All those nodes have to be able to talk to each other in order to fight effectively.

For a typical mission I will sit at a computer console alongside my colleagues and keep an eye on my piece of equipment or piece of the sky. Each mission is a little different, but mostly we scan the sky for enemy communications transmissions and, if necessary, jam them so enemy nodes can't communicate with each other. A mission can last as long as 12 hours, which is a long time to spend in a plane. Our aircraft are big enough to have a few amenities, however, like a microwave and refrigerator. At least we can take coffee-and-doughnut breaks.

I wasn't sure what I wanted to do when I graduated from high school. I knew I wanted to go to college but I had no idea what I wanted to study and, in any case, didn't have the money for tuition. I joined the Air Force to get the GI Bill. I picked it over the other services because it seemed to me like it was more forward-looking and high-tech than the other branches. In fact, the other services make fun of the Air Force because we're not as stuck in tradition as they are. Works for me.

Don't be afraid of basic training. It's hard, but you quickly learn that it's all for a good cause. You'll learn to take orders and work together with your fellow airmen to accomplish difficult and challenging tasks. After a while it actually becomes fun. The Air Force does a great job with training and education. In fact, I've already begun an associate degree program through the Community College of the Air Force, which makes allowances for my hectic travel schedule, and I plan to move on to earn a bachelor's degree as soon as I can. The Air Force is paying for everything. I don't know how long I'll stay in, but I know I started my adult life in the right place."

I Am a Regional Band Apprentice

"Full-time, salaried jobs for musicians are few and far-between. I'm very glad to have one of them. As an Air Force regional band member, I'm learning more about music than I would in any other job because I get to play all day, every day.

I studied music all of my life and have a bachelor's degree in music. My main instrument is the trumpet. Unlike piano or guitar, there are relatively few full-time jobs out there for trumpet players. Most college music students double major in something else, or earn a teaching certificate so they can get a stable job after college. The odds of getting a good full-time job playing your instrument are very slim. In fact, many talented musicians give up music as a profession after they finish college.

I have never wanted to do anything but play my instrument. I auditioned for the Air Force music program and was offered a slot. Best day of my life. Competition for available slots in military music programs is as tough as competition for any job in music, but the prize isn't a short-term gig or a few hours' pay to play backup on somebody else's recording. The prize is a 20-year career during which you can devote all of your time to your music.

The quality of musicianship here is extraordinarily high.

Almost all of the enlisted personnel have at least a bachelor's degree in music, and many have master's degrees. The few officers in this community are all prior-enlisted musicians who earned commissions after many years of service. What makes us different from most musical groups is that we play every day of the week. We work regular daytime hours just like most people, but we spend them all practicing our instruments. Most working musicians have day jobs to supplement their income from making music. Music is my day job.

I really like the military side of our group. We have to take extra-special care of our uniforms because we are always on-stage. And not just on any stage, but on stages with generals, heads of state, foreign dignitaries, you name it. The Air Force band does not play just for entertainment. Music has always been an important part of military tradition, ever since drummers began to liven up the cadences they pounded out to keep soldiers on the march. I take great pride in being a part of that tradition.

There are many opportunities in the Air Force that are not specific to the military. No matter where your interests and talents lie, the Air Force can put you to work, and in an environment with the kind of collegiality that's hard to find in civilian employment. I've been all over the world, doing what I love supporting a cause I believe in. What could be better than that?"

I Am a Food Services Sergeant

"I joined the Air Force to learn a trade. I learned a good one, too. I am a food services sergeant. That means I am a cook and well on my way to becoming a chef. Most people think a chef is just a senior cook. That's partly true, but a chef, first and foremost, is a leader. That's what I want to be: the leader of a fine kitchen staff.

All food services personnel go through three months of initial training after basic training in order to learn how to cook in a high-volume commercial environment. 'High-volume' barely begins to describe the nature of my job. I've been assigned to the main cafeteria at a large Air Force base where we routinely serve four to five thousand meals per day. You think that sounds like a lot? At most bases relatively few people eat breakfast and dinner on base, and there are lots of fast-food places and other options for lunch. Deployment is another matter entirely. I've worked at joint bases with 10,000 personnel from all services at which everybody ate all three meals per day on-base. Oh, and we were on a 12-on/12-off, day-and-night rotation, too. That means we served 30,000 meals per day, around the clock. The logistics of a commercial kitchen are mind-boggling. When we prepare recipes we measure spices by the pound. We order food by the ton. It's something you really have to see to believe.

We also prepare extra-fine meals for VIPs and special events. That's when we get to take our time, measure spices by the teaspoon and hand-select the best meat and produce. When you're serving up dinner for a group of generals you really, really need to get it right.

This job has given me a huge head start on a civilian

career in restaurant management or cooking. I've learned how to do fancy cooking and how to do high-volume cooking. After only four years in the Air Force my skills are sharp and I've had a lot of experiences other cooks will never have. I mean, how many cooks can say they have set up a high-volume kitchen and dining facility in the desert? How about the jungle? Arctic tundra? I've done all that and more.

I'll probably get out of the Air Force after I finish my five-year obligation and use the GI Bill to earn a degree in culinary arts. Because of the skills I learned in the Air Force I'll be able to find a really good part-time restaurant job while I go to school. I have a plan for my life, and it's a good one. I know I have what it takes to make it big in the restaurant business. I can't imagine getting a better start than the one I've gotten from the Air Force."

I Am a C-130 Pilot

"I don't get anywhere near as much attention as the fighter pilots. By comparison to sleek fighter planes, my C-130 cargo aircraft is slow, ungainly and a little ugly. But the C-130 is the workhorse of the American military's ability to move people and assets to wherever they are needed.

Like most pilots, I started out in fighter training. I washed out of the fighter program due to a medical problem that's only a problem when doing the superhuman things that fighter pilots do. So I transitioned to the C-130 training program and have been happily flying the big tubs all over the world ever since.

The C-130 is an amazing aircraft. It's not as large as the C-17, the biggest aircraft we routinely fly, but it's a lot more versatile. The C-17, for example, can only land on real functioning runways. That limits its ability to service really remote areas. By comparison, the C-130 can land just about anywhere. The plane was designed from the ground up to be able to land on grass runways, dirt runways or runways that have been damaged by combat. Its landing gear looks like a cross between tank treads and the wheels on a really big SUV. It's very impressive.

I've taken C-130s into places nobody should ever go. I've flown them through antiaircraft fire, and even been hit a few times. These are tough planes, and they can absorb a surprising amount of damage before they are in any real danger of crashing. Coming back from a mission with a few bullet holes is not uncommon. I've also taken C-130s into jungles and deserts, landing in places nobody else can land. You don't hear enough about it but most of the world's humanitarian aid gets to its recipients via C-130s. We sell C-130s to other countries, and some of those countries use their planes to deliver humanitarian aid, but most people in need rely on the US Air Force to get them food, medicine and professional help. I'd be glad to see other wealthy nations pick up the slack but they won't get very far without investing in the proper equipment. Nobody said C-130s were cheap. Buying and maintaining a fleet requires a national commitment few countries are willing to make.

I love my job because I get a very deep sense of satisfaction from serving my country and helping people in need around the world. You hear a lot about American military 'hyperpower' and how it is allegedly used and abused around the world. My job is a perfect example of how that power can be used to provide assistance nobody else can provide, usually without a shot being fired. And if shots need to be fired, I can supply forward-deployed forces all over the world. I may not get as much attention as a fighter pilot but without me those fighter pilots would have a hard time getting spare parts for their forward-deployed aircraft. That's something to take pride in."

PERSONAL QUALIFICATIONS

NOT EVERYBODY WILL BE HAPPY pursuing a career in the Air Force. If you do not think you can fully dedicate yourself to the Air Force mission of supporting the national military strategy, do not join. The military exists to defend democracy, not to practice it. Nobody will ask you for your opinion until you have some rank and experience. Even then, after the order is given there will be no room for discussion. Everybody has to pull together to make military operations work. If you cannot see yourself really believing in the mission the military is not the right place for you. If you join halfheartedly you will do a disservice to yourself and your country. This does not mean that you have to be politically identical with the commander-in-chief. You do, however, have to have faith in the American system of government. That's why all military personnel take an oath to defend the Constitution. You owe it to yourself and to the people you serve to take this obligation seriously.

It is definitely easier to maintain your enthusiasm for the mission if you can keep the big picture in mind at all times. You may sometimes be called upon to do something you don't want to do and don't think is the right thing to do. The US military is a powerful force in world affairs. It is used to protect the interests of the United States and its allies, and nothing else. It reports to the President of the United States, and nobody else. The President reports to the people of the United States, and nobody else. Many people and institutions would like to turn the US military into something it is not, like a global rescue service to respond to every crisis and tragedy in every corner of the world. Never forget, however, that throughout its existence the US military has been a force for good in the world. Your efforts, even if you may sometimes question them in the short term, are part of a larger strategic philosophy that has, century after century, conflict after conflict, made the world a better place.

If it's sometimes hard to see that your efforts are making much of a contribution, welcome to the military. A bloated bureaucracy under the best of circumstances, the military establishment does not do anything quickly. You must have patience if you want to have a happy career in the Air Force. It is very common to hear service people say that they don't begin to see the fruits of their labors until they near the end of a three-year assignment.

ATTRACTIVE FEATURES

PEOPLE DON'T STAY IN THE AIR FORCE for 20 or more years by accident. They stay because they like it. If you pursue a career in the air force you will be on the receiving end of one of the world's great benefits packages. You will also get your fair share of adventure and have the honor of serving your country.

Nobody ever got rich in the military but the overall compensation package is tough to beat. Job security is excellent and the fringe benefits are exceptional. You and your family will be eligible for medical and dental care provided by Air Force personnel at clinics and hospitals owned and operated by the service. You will have access to commissaries and exchanges not available to civilians. Exchanges use their purchasing power to pass along significant savings to service members, and commissaries (grocery stores) are actually subsidized, resulting in prices that are dramatically lower than in civilian grocery stores. This benefit alone can be worth thousands of dollars a year, especially for a family with children. Housing and moving expenses will all be provided for. If you serve for at least 20 years, you will be eligible for a pension equal to half your base pay for the rest of your life. You will also be able to take military flights free when there are seats available. Very few civilian employers provide the kind of all-inclusive benefits package offered by the military.

For you, adventure will not be something you read about in books or watch in movies. It will be your job. It's popular for military personnel to gripe about being deployed to war zones or to exercises in far off corners of the world. Deployments are never easy and come with many sacrifices, especially for service members who have to leave families behind. But they are also challenging, exciting and, once in a while, even fun. Living in a tent near a flight line in the middle of the desert is hard, but you will find that challenging environments tend to bring out the best in people. You and your colleagues will make the best of the situation and pull together in a way that only happens under duress. You will make memories and friendships that will last a lifetime. Think how many movies have been made about the adventures of military personnel serving in faraway places. Those movies are made because regular people want to experience the adventure that service members take for granted.

Your adventures will not be frivolous, either. You will be living in that tent in the desert because you believe in the greater cause of serving your country. Only about one percent of Americans serve in uniform. They do what they do and take the risks they take so that the other 99 percent can live their secure peaceful lives. There is great pride to be taken in that. In fact, it is the most satisfying part of military service.

UNATTRACTIVE FEATURES

AN AIR FORCE CAREER ISN'T ALL ADVENTURE, however. There is a definite downside. The system can be infuriating. You will have little control over where you go and for how long. Military service can be dangerous.

Like the other military services, the Air Force is an enormous bureaucracy with an extremely comprehensive set of rules. Everything has to be done by the book, every time. Except that it doesn't always work that way. The system is so complicated and unwieldy that everybody bends the rules, but in slightly different ways. Your service record will routinely get screwed up. Every time you move and check into a new command you can expect things to go wrong. No private sector company could operate the way the Department of Defense does. A longstanding joke says that much of military life is about "hurry up and wait." You will be ordered to report to a certain place at a certain time, only to spend hours doing essentially nothing while your paperwork gets shuffled by people you'll never see again. The system will beat you down.

While it's driving you crazy, the system will also circumscribe your life in ways that an ordinary employer never could. When you are on active duty you belong to your service. Evenings and weekends are "liberty" that can be turned into working hours at any time. There is no overtime in the military. Although the system will generally ask you where you want to go when the time comes to change jobs, there is never any guarantee that you will get what you have asked for. Your freedoms of speech and association will be diminished – you will not be able to express all of your opinions publicly, nor will you be able to hang out with whomever you want. Nobody will ask you when you want to be deployed to the desert for six to 12 months. You can never forget that you chose to serve your country, and that service is never about you. The needs of the Air Force will always come first. Although you will always take pride in doing your duty, you may not always be happy about it.

Smart bombs, guided missiles and stealth bombers have led many people to believe that the military isn't very dangerous any more, but that's not true. The military is an unforgiving industrial environment. Just being around aircraft is potentially dangerous. You will find that military bases in general lack many of the comforts typically found in corporate environments. Infrastructure is sometimes not as good as it should be because the military spends most of its money on people and weapons systems. And no matter where you are stationed or what your particular job is, you can always be ordered into harm's way. Ultimately, your job is to risk your life to protect others. That is what you signed up to do and it's only a matter of time until you'll have to do it.

EDUCATION AND TRAINING

THE AVERAGE EDUCATIONAL LEVEL OF military service members is much higher than that of the civilian population. Officer candidates are required to have earned bachelor's degrees in order to be considered for commissions. With few exceptions, only five percent of new enlistees can be high school dropouts. Dropouts are required to earn a General Equivalency Diploma while they are in basic training. This only makes basic training even harder than it already is. If you plan to enlist, plan to graduate from high school first.

If you intend to apply for a commission you will need to earn a bachelor's degree. Although you do not have to major in anything specific in order to meet the minimum requirement of a bachelor's degree, not all degrees will qualify you for all career paths. A degree in philosophy or history may be sufficient if you want to become an intelligence officer, but it won't help you become an engineering officer. The Air Force, like most employers, wants its engineers to have at least a Bachelor of Science degree in engineering before they are hired. Many pilots have degrees in aeronautical engineering before they join. Quite a few have earned private pilot's licenses. If you want to be an officer, choose your major wisely. If you want to seal the deal early, be sure to go to a university with an Air Force Reserve Officer Training Corps, or ROTC.

Be prepared for additional education and training whether you enlist or pursue a commission. The US military is essentially the world's largest continuing education program. Training is constant and ranges from simple, computer-based exercises that can be completed in less than an hour to field exercises that can involve multiple services and even other countries and can last for six months or more.

Officers, in particular, will be required to earn multiple professional certifications throughout their careers. Many officer career paths require master's degrees. If that's the case, the Air Force will provide you with the means to earn one. You will attend either a civilian university at Air Force expense or one of several universities operated by the Department of Defense, like the Air University or the National War College. Enlisted personnel can take advantage of the GI Bill and Tuition Assistance to earn college degrees while they serve. The military provides ample opportunities for education and personal enrichment.

No matter what level of education you achieve before you join the military, you will start your career with initial accession training. Enlistees attend eight-and-a-half weeks of Basic Military Training, commonly known simply as "basic," while officer candidates attend 13 weeks of Basic Officer Training, commonly known as OTS, the initials for Officer Training School at Maxwell Air Force Base in Montgomery, Alabama. Both programs are challenging and will teach you things about yourself that you never knew. Most military personnel look back fondly upon basic training even though almost nobody would volunteer to do it again.

You may be nervous about getting through basic training. This is mostly because of movies that depict basic as a lethal regimen of screaming drill instructors and impossible peer pressure. Basic training is not supposed to be easy, but you will not be hounded to within an inch of your life by a crazed drill sergeant. Modern basic training is a largely academic exercise conducted mostly in classrooms. The physical requirements are challenging, to be sure, but they are not superhuman. If you do some training before you go – which you should – you will be fine. Just think of basic or OTS as college courses with unusually demanding gym classes. Do not let an irrational fear of basic put you off.

After basic training or OTS, you will attend either an enlisted technical training school or officer follow-on training to learn a specific skill. These training programs vary in length and scope, but they all will set you on your career path in the service. You will receive your first assignment after you complete your initial accession training. All of these steps can be negotiated in your enlistment contract. You may not know where your first assignment will be but everything up to that point should be negotiable and reflected in your contract.

Air Force Academy

Another option is to enroll in the Air Force Academy. Like its counterparts operated by the other services, the United States Air

Force Academy in Colorado Springs is an elite four-year university and has equally rigorous admissions requirements. Applicants are assessed based on their grades, test scores and extracurricular achievements, just like any other top ranked university, but also on their community involvement, physical fitness and leadership ability. Applicants also need a recommendation from their US senator or representative. Cadets attend school almost year-round for four years and are obligated to serve for at least eight years after graduation, consisting of five years of active duty and three years of reserve duty. If you think you have what it takes you can learn more at this website:

www.academyadmissions.com

EARNINGS

CONVENTIONAL WISDOM SAYS THAT nobody gets rich in the military. That may be true but the military's overall compensation package is hard to beat. Salaries are quite competitive with civilian occupations requiring similar education and experience. The military offers additional compensation, such as a housing allowance and a cost-of-living allowance, which makes the deal even better. It should be noted that the military compensation package is also intended to take care of the spouses and children of service members. Personnel are more comfortable about deploying for long periods of time when they know their families will be taken care of while they are away.

Enlisted Service members

In all military services, your paycheck is actually composed of several different kinds of compensation. The largest portion of your paycheck is base pay. This is your salary without benefits. A new enlisted airman in the rank of E-1 (airman basic) currently is paid $1,400 per month, or $16,800 per year. This may not sound like much, but keep in mind that most E-1s are in basic training or initial skill training where they also get housing, three meals a day and almost no opportunities to spend money. Base pay jumps to $1,569 per month for E-2s (airman) and $1,588 for E-3s (airman first class).

As a young airman or officer, most of your needs will be provided for, so your smallish paycheck will actually go pretty far. When you rise to E-4 (senior airman) your base pay will jump to $1,921 per month if you've been in for at least two years or $2,025 per month if you've been in for three years. An E-5 (staff sergeant) with four years earns $2,335 per month at current rates. After that, your base pay will increase every two years all the way up to $5,928 per month for an E-9 (chief master sergeant) with at least 26 years of service. The highest-ranking personnel can, under certain circumstances, serve for as long as 40 years. With a combination of hard work and luck you could top out the pay scale at $6,830 per month in base pay for an E-9 with at least 38 years of service.

At some point you will probably move off base and into a house or apartment of your own. When you do you will be entitled to a housing allowance. Housing allowances are calculated based on your rank, location and whether or not you have dependents. The housing allowance for a single E-5 airman living in a rural area, for example, would be substantially less than the housing allowance for a married E-5 airman living in an expensive metropolitan area. Housing allowances for some areas, especially overseas, can be as large as or larger than your base pay. Typically, however, a housing allowance will equal about one third of your total earnings.

After your base pay and housing allowance, there are additional entitlements. The cost of living allowance, or COLA, adds to your base pay when you are assigned to an area with a high cost of living. Hazardous duty pay adds additional money when you deploy to a dangerous area. Your paycheck will be tax-free if you are deployed to an area declared a combat zone by Congress. That is a substantial raise. Airmen who can prove their fluency in certain languages may be eligible for a monthly bonus. The list of ways to boost your paycheck is quite lengthy. All of these bonuses will require hard work on your part. Want to earn an extra $600 per month? Learn Pashto, the language spoken in Afghanistan – that's really hard.

Officers Earnings

Pay scales for officers are higher than those for enlisted airmen. Most other allowances are higher too. Base pay for a new O-1 (2nd lieutenant) starts at $2,655 per month and rises to $3,483 for an O-2 (1st lieutenant) with two years of service. From there the pay scale rises to a maximum of $18,061 per month for an O-10 (general) with at least 38 years of service. Keep in mind, however, that there are only about a dozen four-star generals serving at any given time.

OPPORTUNITIES

YOU HAVE MANY OPPORTUNITIES TO learn more about careers in the Air Force before you make your move. You can join the Reserve Officer Training Corps, or ROTC. You can also take an intermediate path and join the Air Force Reserve or Air National Guard.

ROTC

If you plan to join the Air Force as an officer you will have to earn a bachelor's degree. Since you will be going to college it will behoove you to seek out a university that offers an Air Force ROTC program. In exchange for a full or partial scholarship, you will be required to serve for at least four years after you graduate. Pilots are required to serve for at least 10 years as a result of the expense of their training. A few other career paths, like navigators, are required to serve for six years. ROTC is a minor concentration in military science taught by active-duty Air Force personnel. ROTC cadets typically wear their uniforms on campus one day per week, spend several weekends each year doing exciting things few other students get to do, and go on a challenging two-week assignment every summer. ROTC cadets are offered commissions upon graduation. If you have your sights set on becoming a commissioned officer, ROTC is an excellent path to take. In fact, ROTC is the Air Force's single-largest commissioning source, ahead of OTS and the academy.

Air Force Reserve

The Air Force also offers an intermediate path for those who want to pursue civilian careers while serving their country. The Air Force Reserve offers part-time careers. Initial accession requirements for the reserve are the same as for active-duty careerists. After completing them, however, reservists are obligated to serve only for one weekend per month and two weeks per year. Reservists can volunteer to fill active-duty positions for anywhere from one month to two years, which can be an exciting alternative to a civilian job. Reservists can be called up to augment active-duty forces at any time. Such involuntary activations can last from one to three years. Many reservists cultivate rewarding part-time military careers that complement their full-time civilian careers. Others turn their reserve careers into something as far removed as possible from their civilian careers, for the sake of variety in their lives.

Air Force National Guard

The Air National Guard presents similar opportunities. Incorporated as state militia, all 50 states have Air National Guard organizations. Initial accession requirements are the same as for the active and reserve components. After completing training, however, guardsmen and guardswomen are typically posted to the unit nearest to their home. It is possible to spend a 20-year guard career in the same unit. National Guard members are required to serve for one weekend per month and two weeks per year, and can be called to active duty like reservists. The National Guard maintains a small force of full-timers to run its operations, and members routinely step up for a year or two to fill these positions. Although affiliated with the Air Force, the Air National Guard is a legal entity unto itself. Typically under the command of state governors, National Guard units fall under the purview of the President only when they are deployed outside the United States. Regular military components are prohibited by law from engaging in law-enforcement activities within the United States. National Guard units, however, can be deployed by governors to augment law enforcement during times of crisis.

GETTING STARTED

IF YOU HAVE DONE YOUR RESEARCH, thought about your options, and decide you want to pursue a career in the Air Force there's really only one thing left to do: Talk to a recruiter.

Talking to a recruiter does not impose an obligation to serve in the Air Force. In fact, most people who join the military talk to their recruiter many times before signing the final contract. It is common for recruiters to meet with parents and guardians, too. Nobody will seriously suggest that you rush blindly into such a big decision. The Air Force is a continuum of people, however. That means that there are people constantly moving through the career cycle, with people joining at one end and getting out at the other end. Because of this, the needs of the service change constantly. A recruiting station is the only place to get up-to-date information on current opportunities.

Part of determining your career path in the Air Force requires taking a test. If you plan to enlist you will have to take the Armed Services Vocational Aptitude Battery, or ASVAB. A standardized test not unlike the SAT or ACT, the ASVAB measures your aptitude in several areas related to military service. Your score will determine which jobs you qualify for. The higher your score, the more opportunities you will have. Officer candidates are required to take the Air Force Officer Qualifying Test, or AFOQT. Taking the test does not incur any obligation. If your score doesn't give you the opportunities you're looking for, you don't have to join.

Whatever you do, do not sign on the bottom line until you have negotiated the deal you want. Your enlistment contract should spell out your initial accession requirements and the training programs you will be required to complete on your way to the career path you want to pursue. Do not sign anything until your contract says what you want it to say. Remember, you are under no obligation until you say you are. When that day comes, go for it! There is a great adventure waiting for you. Good luck.

ASSOCIATIONS, PERIODICALS, WEBSITES

☐African Center for Strategic Studies
www.africacenter.org

☐Air Force Link
www.af.mil

☐Air Force Outreach
www.afoutreach.af.mil

☐Air Force Reserve Command
www.afrc.af.mil

☐Air Force Reserve Officers Training Corps
www.afrotc.com

☐Air Force Times
www.airforcetimes.com

☐Air Forces Monthly
www.airforcesmonthly.com

☐Air International Magazine
www.airinternational.com

☐Airman Magazine
www.airmanonline.af.mil

☐Air National Guard
www.ang.af.mil

☐American Legion
www.legion.org

☐BAE Systems

www.baesystems.com

☐Boeing
www.boeing.com

☐Canadian Air Force
www.airforce.dnd.ca/site/index_e.asp

☐Centre for Air Power Studies
www.airpowerstudies.co.uk

☐China Defense Today
www.sinodefence.com

☐Combat Aircraft Magazine
www.combataircraft.com

☐Combined Joint Task Force Horn of Africa
www.hoa.africom.mil

☐Dassault Aviation
www.dassault-aviation.com

☐Defense Link
www.defenselink.mil

☐Defense News
www.defensenews.com

☐Flight Journal
www.flightjournal.com

☐French Air Force
www.defense.gouv.fr/air_uk

☐French Foreign Legion
www.legion-recrute.com/en/

☐Global Security
www.globalsecurity.org

☐Israeli Defense Forces
www.dover.idf.il/IDF/English/

☐Jane's
www.janes.com

☐Joint Strike Fighter
www.jsf.mil

☐Lockheed Martin
www.lockheedmartin.com

☐Marshall Center for European Center for
Security Studies
www.marshallcenter.org

☐Military.com
www.military.com

☐National Museum of the Air Force
www.nationalmuseum.af.mil

☐North Atlantic Treaty Organization
www.nato.int

☐Northrop Grumman
www.northropgrumman.com

☐Royal Air Force
www.raf.mod.uk

☐Royal Australian Air Force
www.airforce.gov.au

☐Royal New Zealand Air Force
www.airforce.mil.nz

☐Scramble
www.scramble.nl/airforces.htm

☐Swedish Air Force
www.mil.se/sv/I-Sverige/Flygvapnet/

☐Today's Military
www.todaysmilitary.com

☐**United Nations**
www.un.org

☐**United Nations Security Council**
www.un.org/docs/sc

☐**United States Air Force**
www.af.mil

☐**United States Army**
www.army.mil

☐**United States Coast Guard**
www.uscg.mil

☐**United States Department of Defense**
www.defenselink.mil

☐**United States Department of Veterans Affairs**
www.va.gov

☐**United States Institute of Peace**
www.usip.org

☐**United States Marine Corps**
www.marines.mil

☐**United States Naval Institute**
www.usni.org

☐**United States Navy**
www.navy.mil

☐**Veterans of Foreign Wars**
www.vfw.org

Made in the USA
Middletown, DE
28 December 2018